Digital Self Mastery Across Generations

*How to Master Your Relationship with Technology
to Amplify Productivity
and Connection in the Digital Era*

by

Heidi Forbes Öste, PhD

Download a free copy of the tips & exercises from
this book at 2BalanceU.com

Legal And Copyright

Digital Self Mastery Across Generations: How to Master Your Relationship with Technology to Amplify Productivity and Connection in the Digital Era

Copyright ©2018 Heidi Forbes Öste, PhD

All rights reserved. No part of this publication may be reproduced, distributed, or transmitted in any form or by any means, including photocopying, recording, or other electronic or mechanical methods, without the prior written permission of the publisher, except in the case of brief quotations embodied in critical reviews and certain other noncommercial uses permitted by copyright law.
2BalanceU.com

info@2BalanceU.com
San Francisco, CA Boston, MA & Malmö, Sweden

2018 Version 1.2

Ordering Information:
Quantity sales. Special discounts are available on quantity purchases by corporations, associations, and others. For details, contact the publisher at the address above.

Printed in the United States of America

Limit of Liability Disclaimer: The information contained in this book is for information purposes only, and may not apply to your situation. The author, publisher, distributor and provider provide no warranty about the content or accuracy of content enclosed. Information provided is subjective. Keep this in mind when reviewing this guide.

Neither the Publisher nor Authors shall be liable for any loss of profit or any other commercial damages resulting from use of this guide. All links are for information purposes only and are not warranted for content, accuracy, or any other implied or explicit purpose.

Earnings Disclaimer: All income examples in this book are just that – examples. They are not intended to represent or guarantee that everyone will achieve the same results. You understand that each individual's success will be determined by his or her desire, dedication, background, effort and motivation to work. There is no guarantee you will duplicate any of the results stated here. You recognize any business endeavors has inherent risk or loss of capital.

Note from the Author

Digital Self Mastery Across Generations is written as a foundation piece to prepare you for deeper individual. I hope that it can provide guidance and food for thought for each of you as you engage with technology and allow your digital self to evolve.

This book is the second in the Digital Self Mastery series. If you read the Online Entrepreneurs' edition (the first), Part One may be familiar. That said, this edition is written from a different perspective, and in language that aims to reach a broader audience. It also presents new insights and updated information and references. The simplified theoretical review provides context for the digital self.

Digital Self Mastery Across Generations provides a fresh angle from which to look at your relationship with technology. I hope you enjoy it. I look forward to participating in an ongoing conversation with you.

You will note that there are exercises throughout this edition. They are designed to help you understand not only where you and others fit on the digital self spectrum, but also how to shift to a more balanced digital self. Please write in this book. Use the white spaces to jot down your thoughts.

Since publishing the Digital Self Mastery series, I have become more active in advocating for Digital Wellbeing & ethical use of behavioral science in the development and design of new technologies. You can learn more about this by subscribing to the Evolving Digital Self podcast (available on most podcast apps).

On a personal note:

I would like to express my sincere appreciation for those who have provided the necessary support, from heart and head, in the process of creating Digital Self Mastery Across Generations. It has been my intention to create a tool that can help bridge the gap between technology and humanity. It is a challenge that I gladly undertake.

This edition expresses Digital Self Mastery in a far more user friendly, if you will, non-academic "blah blah blah" manner. Thanks to my editors, Sam Eddy and Tally Forbes for their painstaking efforts to proof and review the manuscript. Mike and Ed, for their mentorship and belief in my ability to "get out of my head." It opened my heart to the place where it needed to be in order to share my work with a wider audience, like you.

Thank you to my children, Oskar and Hanna-Maria, and my parents. They put up with my persistent questions to be sure that I represented their generations more accurately. As a result, their comments helped me understand my own generational bias and how to communicate it to you, the reader. More importantly, they provided me inspiration to see that the future that they represent is bright.

Thanks to my love, Björn, for just be-ing.

Contents

Introduction — 6

Part One: The Why... — 13

- Chapter 1. The New Norm — 14
- Chapter 2. Social Optimization — 29
- Chapter 3. Making Peace — 43
- Chapter 4. Digital Self — 53

Part Two: The Who... — 67

- Chapter 5. Digital Averse — 69
- Chapter 6. Digital Resistant — 76
- Chapter 7. Digital Cautious — 85
- Chapter 8. Digital Balanced — 90
- Chapter 9. Digital Curious — 97
- Chapter 10. Digital Hoarder — 101
- Chapter 11. Digital Addict — 106

Part Three: The Where to... — 112

- Chapter 12. Evolving — 113
- Glossary — 119
- Notes and Doodle Space — 127

Introduction

From smart phones to smart homes, in the digital era, technology is pervasive in all areas of your life. Professionally, unhealthy relationships and poor habits with technology can determine the difference between struggling to manage your time and energy and having the life that you desire and deserve. Personally, those habits and unconscious behaviors can impact your relationships with friends and family, your sense of self and, more importantly, your general wellbeing.

Today, a greater part of your social and business interactions take place through technology in some form. In order to evolve and thrive in the digital era, you will need a greater understanding of your relationship with technology. Not only yours, but also that relationship for the people in your life and in your work. Without it, you can expect an environment overflowing with miscommunication and frustration.

You may be exploring this as an individual, leader, or even parent. Regardless of why you were drawn to this topic, your curiosity is the first step in recognizing the impact of this relationship and how it varies for different people in your life and work. Let's start with the big issue and we will work down to the small- what you can do or change.

Both individuals and organizations are increasingly focused on improving wellbeing. The data shows that wellbeing impacts productivity, engagement and turn over. Ultimately, wellbeing has a good return on investment.

This definition often includes physical, psychological and financial wellbeing. What is missing from the equation, as it impacts all the above as a delivery mechanism and a factor introducing its own challenges, is the inclusion of *digital wellbeing*.

Digital wellbeing is has three elements:

1. The first element is ethical use of behavioral science when developing or choosing new technologies and impact on general wellbeing.

2. The second element is designing products and services that support rather than sabotage human wellbeing.

3. The third and most relevant to the individual and digital self, is building healthy boundaries to achieve a less toxic relationship with technology.

Digital wellbeing strategies integrate better practices and behaviors that support a positive relationship with technology for both users (like you) and developers of new technologies. Without integrating digital wellbeing strategies, productivity and engagement can be damaged. Physical, mental and financial wellbeing are also greatly impacted. Without digital wellbeing, the price you pay is high. You diminish your most valued assets; your health, human connection and time.

Without the integration of digital wellbeing practices, negative patterns can develop that are destructive to both individual (you) and even your organization's potential growth and relevance. These patterns are related to old behaviors that have not adapted to the new and rapidly changing tools and environments. Change can be challenging for anyone, and digital transformation is accelerating quickly.

Hold onto your hat, because you will find there are some old habits that will come in handy to survive this ride.

Learn how to not only survive but also thrive at integrating pro-active digital wellbeing & ethics strategies. You will find that both the products and services that you use, or if you are a designer/developer that you create, integrate better into the human experience. Your solutions will have more lasting and sustainable outcomes. They will work with you rather than against you.

Technology solutions are improving the ability to understand your body and your mind. Daily interactions with others are increasingly mediated or at least initiated through some form of technology. Yet, you place little importance on the relationship with this technology that has become so ubiquitous in your life and your work.

Society today is quick to blame technology for the broken social systems and anything else that doesn't work the way we feel it should. Technology, itself, is neutral. Technology is the megaphone or amplifier. It is important to recognize the source. The human (bad, good or apathetic) designing or using it has enhanced reach and impact with technology as their amplifier.

That said, in any intimate relationship, if you don't take time to recognize, appreciate and understand its value and its limitations, the relationship has the ability to grow toxic. From averse to addictive behaviors, you are now more than ever in need of creating a balanced relationship with technology. Technology is not going away. In fact, as technology becomes more and more integrated in the processes for creating sustainable futures, you need to take ownership of your role in this relationship in order for it to become the powerful tool it was intended to be. The goal of technology is to make you better, more efficient, healthier and happier.

Digital Self Mastery integrates human development and behavior science with tech- savvy strategic business and systems thinking. As the digital becomes more integrated in your life and work, your relationship with it becomes ever more critical to manage. Digital Self Mastery Across Generations presents new solutions for the rising challenges of the digital era across the generations in the workforce and at home.

Weaving real life stories from leaders and practitioners around the globe, Digital Self Mastery is essential for humans and organizations to thrive.

Note: There is a glossary at the end. Don't be afraid of words with "new" meanings. Enjoy!

Part One:

The Why...

Chapter 1. The New Norm

> **anomaly** | *noun* | anom·a·ly \ə-ˈnä-mə-lē\ : something different, abnormal, peculiar, or not easily classified

The Anomaly

The anomaly is the norm. You are defined by more than your DNA. Your stories are rich with experiences and relationships that create both your potential and hidden barriers. They refine who you are as an individual and how you communicate with your environment.

In the digital era, your experience is expanded exponentially both in time and space. You can now simultaneously interact with clients and peers across the globe in multiple languages without having to leave your home or learn a new language. Your differences are your strengths. Your ability to participate in interactions mediated by technology limits isolation and binds us all together as a community. Removing the barriers to participation, often triggered by beliefs or experience, opens you to the richness of perspective and the opportunities of the global marketplace and community.

Digital Self Mastery is a culmination of my experience and appreciation of the benefits of twenty-five years in social strategy practice as a consultant, a deep scholarly dive into the behavioral science behind the human experience with technology, and a personal passion for people, connection, systems, and wellbeing.

I write this from the lens of a practitioner, scholar, mother and most importantly a fellow human. I love technology and the opportunities that it affords me when I am in a good relationship with it. I am most comfortable when out of my comfort zone. Whether you read this as a technologist trying to understand the human side of the relationship or the counter. My role has been, and still is, to act as a bridge between technology and humans. I, like you, am an anomaly.

As a social strategy consultant, I developed a process for *social optimization* working with individuals and organizations. This process was used to teach the benefits of using social technologies to build and maintain MUTUALLY beneficial and effective relationships. Relationships are increasingly critical, as is mastering the blend of physical and virtual interactions. Awareness of teams' relationship with technology is just as critical as your own for effective collaboration and connection.

As a scholar, I expanded on this work to develop *social optimization theory*. (You will learn more about this in the next chapter.) Presence, being truly in the moment, is critical to achieve the reciprocity (mutual) factor used to evolve. Human connection and being consciously present is a growing issue. Among my research findings was the impact of the human relationship with technology on the ability to benefit from the intended outcomes of new technologies.

As a mother of GenZ (born just after 2000) teens, an entrepreneur, expat, member of a large extended family and wife of a Swedish entrepreneur, a harmonious relationship with technology is key. I rely upon and appreciate the benefits that digital self mastery affords. Through connection, improved productivity, efficiency and a sense of being part of the greater global system, we each provide our unique value to the system when we master it.

As a Digital Wellbeing & Ethics advisor, I support the design, development and launch of technologies that enhance the human experience rather than becoming it. I am working from all angles to improve your relationship with technology from its conception to implementation to your interaction with it and its even end of its lifecycle. My aim for you is that this work will transform your relationship with technology from one that works against you to one that works for you.

Harnessing the power of the digital self make the difference between the user's failure and isolation or success and engagement. Building self-awareness of your interactions with the changing landscape and what they afford you is just the beginning. Your unique stories and paths, both the chosen and imposed, influence your behavior and response. You can choose to harness this, engage in it or be overwhelmed and anxious.

Presence with yourself, others and the system around you is the ultimate goal.

Three Influencers

In *Digital Self Mastery Across Generations*, variations in the relationship with technology are influenced by three things: when it was introduced into your life, your life stage, and your generational culture. In my interviews for this book, I discovered that the variation was less about the assumed generational differences and more about life stages, personality and the unique characteristics and story or experiences of each individual.

The anomaly is the norm. The increasingly more customizable and digitized world supports the variability in each human experience and story. The technology you use becomes part of your self-expression, function and maintenance. No longer are you simply referring to the ubiquitous mobile phone. The device you carry is your wallet, alert, lifeline, communications, coach, organizer/calendar, ticket agency, doctor, memory, public relations team, and the list goes on.

How you express your unique self in the changing digital landscape depends greatly on your relationship with your technology.

Some gravitate to the comfort of sameness and the ability to be a silent observer. Others feel the ability to personalize is the ultimate in control and peace of mind. What remains the same is that we are all different and the changing ecosystem provides opportunity for the anomaly in all of us.

The Human Factor

To be yourself in a world that is constantly trying to make you something else is the greatest accomplishment.

Ralph Waldo Emerson

What do Nilofer Merchant's New Social Era, Seth Godin's Tribes, Yuval Noah Harari's Sapiens, Stephen Pinker's Enlightenment Now all have in common? The human factor of doing business today and the ability technology affords you to connect-regardless of time and space. By including the human factor in the connection that technology creates, you develop wider reach with deeper and more effective relationships that sustain your businesses and communities.

This new paradigm creates an environment where we are dependent on high levels of social intelligence, which goes beyond the need for high IQ, empathy and emotional intelligence. The shift requires the ability to develop relationships based on social awareness and your facility to engage authentically in those relationships.

The wellbeing of organizations, their people and their stakeholders is a factor that weighs in heavily in this new paradigm. If we are not well, we miss and are not alerted to social cues, not only of others' but also our own.

The human element of relationships is challenged by connecting through mediated tools; phones, video conferencing, messaging, texting, and even games.

Organizations invest over $500M each year in personality, aptitude and compatibility testing of their employees, teams and students. They are missing the fact that technology is the means for many of the interactions- it is either a conduit or a bottleneck. Digital Self Mastery aims to provide a greater understanding of this relationship in order to diminish the bottlenecks and misinterpretation that disrupt vital connection of human relationships.

Presence is key. The most valued human connections are obtained through trust, which is strengthened by being fully present. *Presenteeism* is when you are "checked-out" at work but still show up. Technically it is the lost productivity as a result of physically being present without presence of mind. In my 2015 dissertation and eventual book, *BE-ing@Work*, I enhanced the definition of presenteeism to include disengagement and cognitive (or unconscious) to the original definition of health-related disruptions to presence.

When the relationship with technology is toxic, ranging from aversion to addiction, it results in disengagement presenteeism. It is important to recognize this type of presenteeism, along with the traditionally defined health-related and cognitive, in order to develop solutions for improving the human connection and productivity in the workplace, school and at home.

We are increasingly aware of the interconnectedness and impact of physical and mental wellbeing. There is a growing movement towards conscious business practices.

This is seen in the rise of B Corps (using business as a source for good, ie. Ben & Jerry's and Patagonia), the growth of mindfulness in organizations (ie. Google, Siemans and General Mills) as well as more social enterprise models (ie. Skoll Foundation & Branson's B Team).

These are just some examples of this shift. On a smaller scale is the movement of individual entrepreneurs (ie. Tony Robbins, Mike Koenigs) who build balanced giving into their business model. These models support people, planet and profit and recognize the humanity behind and within the machines that run our businesses today. If you can use technology to accomplish that, going forward, everyone wins.

The Digital Era

People need to understand how exponential technologies are impacting the business landscape. They need to do some future- casting and look at how industries are evolving and being transformed.

- Peter Diamandis

Before you read this chapter, take a moment to observe your surroundings.

- * Can you see anything that is not touched by technology and innovation?
- * Are you aware of how technology is integrated into your life?
- * What is the effect on your quality of life and ability to thrive in life and business?

Technology touches nearly everything you have today and is based on co-creation. You can chose to deny the value it creates and fight it, or you can embrace the Digital Era, master your own relationship with it and thrive. It is a choice, but it does require active engagement.

- * If a person did everything that your technology does, would you have a different appreciation for them?
- * Would that person be your best friend and most appreciated ally, or would you resent them for your

dependence upon them? I would hope the former.

Futurists like Brian Solis and Gerd Leonhard embrace the potential of machine learning and artificial intelligence (AI) to improve and evolve beyond our human capacity. Machines extend human capacity for efficiency and productivity. That said, humans have the unique additional capacity and requirement for emotional and social intelligence.

Let me emphasize this again, as it is important for you to remember.

> **HUMANS** have the unique additional capacity and requirement for emotional and social intelligence.

As Gerd Leonhard says, "efficiency is for robots, but happiness is the final destination." But as Solis points out in the *Future of Work*, "in a world of AI, machines, and robots, humanity will be the killer app." In a recent conversation with Solis on Evolving Digital Self, he emphasized *digital darwinism* as a culture and organizational change issue that cannot be ignored.

Technology is not limited to smartphones, although they do receive the greatest attention (both in conversations around the relationship with technology and from us physically). Digital health has expanded beyond the pacemaker to provide solutions for previously debilitating conditions, remote treatment, and greater accessibility. Wearable technologies provide incredible insight into our personal health and wellness data. Some, like the AppleWatch, even become extension of, or even replace, the smartphone.

You may notice my personal passion for digital health and wellbeing devices. This is not simply because of my dissertation research on *wellness wearables and presence of mind in the workplace*. It began with my use of wearable fitness trackers to manage my physical and mental wellbeing while suffering from debilitating Lyme's Disease during graduate school.

My passion for wearables became cemented when my AppleWatch (which came out after my study) saved my life. The passive tracking of my heart-rate enabled my doctor to identify the spread of Babesia (a co-infection of Lyme) into my heart and provide immediate treatment remotely. This is just one of many incidents in which I was thankful for this amazing piece of technology attached to my wrist.

The vast array of features and functions that are intended for peace of mind and presence, but require digital interaction is increasing. There is fast growth in mindfulness technology specifically designed to help you reconnect with yourself and others. Breathe, Whil, Calm and LiveAMoment are examples that teach you not only to pause but also to acknowledge how that feels. Smart home and car features link you to the ability to monitor and optimize your environments for better efficiency, safety and comfort. These innovations are designed to improve your quality of life and touch all industries, thereby changing the way you do things. Your goal should be to learn how to integrate them peacefully.

Whether you like it or not (let's hope I can help you learn to like it), the rapid evolution of technology, and its integration into all aspects of our lives and worlds, is the basis for the digital era. In the last century, the capacity, complexity, and power has followed Moore's Law to exponentially increase while access, cost, size and adoption periods have reduced.

The implication is that to operate this supercomputer that once would has filled an entire room and required a team of highly trained professionals, a simple voice command to a pocket sized device will now suffice. Even better, with connected devices like Amazon's Alexa, Apple's Siri and Google's Home even a small child can operate it.

This was demonstrated to me when visiting a dear friend last winter. I was struck by his three-year old's excitement at commanding Alexa to play, "Happy." The squeal and wiggles that followed even made us dance. I must admit I had a similar level of excitement (seriously) upon asking Alexa to "play Digital Self podcast" for the first time. Oh, the joys of hands-free control!

Since publishing the first Digital Self Mastery, there has been a mounting backlash towards technology, in particular social media. Several high-level executives and engineers at major tech companies in Silicon Valley have spoken out about design for addiction, manipulation of democracy, influence on mental health, social interaction and even children. This has sparked the development of a movement, originally called *Time Well Spent* by Tristan Harris (former Design Ethicist at Google), now a consortium called Center for Humane Technology.

I support their efforts 100%, but also wish to clarify where this book fits, in context with their work. The backlash is predominantly directed at social media, which I must emphasize is not what this book is about, as it is only one element of technology. In recent arguments technology is often generalized when referring to the evils of social media and/or the smartphone.

The reality is that technology is far more integrated into our lives and work. Smartphones are used for more than social media. Gaming strategies can be used for the greater good by developing positive behaviors. Basically, there is a lot that we can gain from developing a positive relationship with technology.

Digital Self Mastery is intended as an alternative perspective to understanding the relationship profiles between humans and technology. It also explores how to shift that relationship to achieve better balance and thus enhance your life and that of others. The digital era is not going away anytime soon, so you might as well learn to master it in your own way.

In the next chapter, try to identify where and when you

evolved from one stage to the next of social optimization. What might have been the trigger for you to either rise or fall? This will help put things in context as you move forward in mastering your digital self. Your experience is unique, and therefore where your journey leads will also. Embrace your anomaly.

Chapter 2. Social Optimization

> **Social Optimization:** the building and maintaining of **MUTUALLY** beneficial and effective relationships

Social Optimization

I live and breathe the concepts behind *social optimization* in my work and life, and have for many years. In 2008, this became the foundation for my work in teaching the Art of Social Strategy.

Please bear with me here as I geek out a little. I promise to keep this part short, but it will help you understand the ups and downs of the evolution of your relationship with technology.

In this chapter we will explore the evolutionary process as it applies to your relationship with technology. In the relationship economy, discussed in the previous chapter, social optimization creates the bridge to reciprocity and building trust. Even initial face-to-face connections are sustained and even expanded when separated by time and/or space. Technology mediates our interactions, either initially or in maintaining relationships. As a direct effect of rapid adoption of new social technologies both in the workplace and interpersonal communication, we need to be at peace with the technology in order for relationships to thrive in this heavily mediated environment.

Social technologies in this context are any technology that enable interactions between ourselves, others and the systems: smart phones, social media, wearable tech, tablets, social gaming, augmented reality, music sharing, social shopping, social search, location based services, etc... By developing visuals, my aim is to provide clarity to a complex concept. Just because this is based on scholarly theory, should not mean it's relevance is only for scholars. After all, that would not be applying the principals of social optimization and walking my talk.

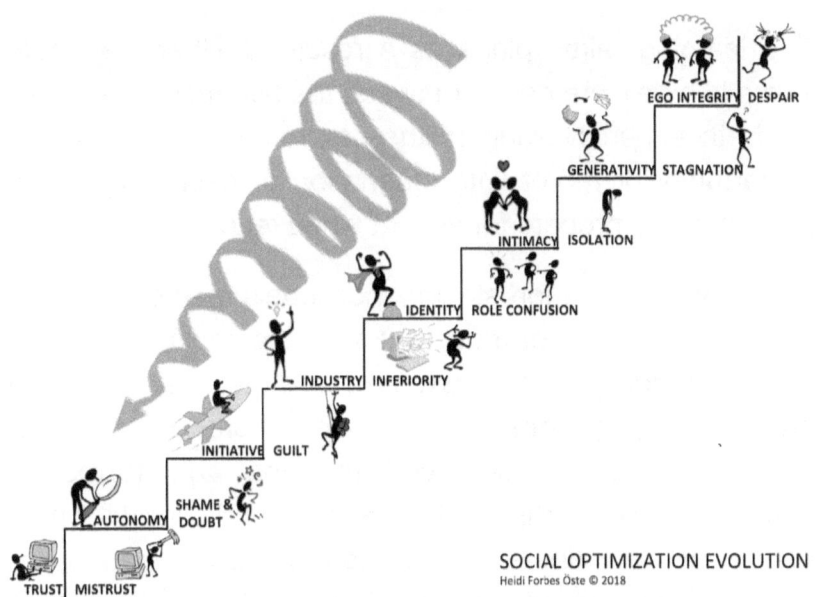

SOCIAL OPTIMIZATION EVOLUTION
Heidi Forbes Öste © 2018

The Evolution

Evolution from one stage to the next is a result of *kairos* (timely and appropriate opportunity) or crisis. Depending on the response, it can lead to *metanoia* (transformative change of heart or mind) or evolutionary development through the stages.

The learning takes place as a result in either response. There is a delicate dance between the two responses. This leads to either evolving to the next stage by seizing the opportunity or regret with stagnation or descending to a prior stage to prepare for ascent when ready.

In Social Optimization stages, individuals may descend or ascend as a result of triggers (kairos/metanoia) that occur in the changing *tech ecosystem*, like the shift to mobile and messaging apps. (I will touch more on the tech ecosystem later.) New technologies and forms of communication abruptly change the conditions in which you exist and operate and therefore your stage. For example, you may move from *Generativity*, at which point your focus is on creating a legacy beyond yourself, to *Industry*, in order to re- establish your expertise in the new conditions.

Depending on whether you embrace the new industry or resist it, you will rise, remain at your new stage or even descend. Upon descending, you will have to rise again in the new conditions, which are ever changing. Just as in your first ascent, you cannot skip stages as the conditions in the new paradigm have changed. That said, you may ascend faster, as a result of the learning in the previous ascension process.

To keep it simple and be sure that each of the stages is relevant, I broke it down for. Each stage correlates with the mutually beneficial relationship between humans and the technology required to evolve to the next stage.

Learning the technology does not inherently mean that you can apply it in the context of interpersonal connection. *Digital natives,* those born into an era and an environment where computers were prevalent, are not necessarily pre-disposed to achieving the highest level of social optimization.

This model is loosely based on Erikson's psychosocial stages. Just because one can teach stages intellectually does not mean that they are at the highest stage. Disconnect between the intellectual understanding and the personal, moral, ethical development is a shared concern in both models. As we will discuss in chapter four on the digital self profiles, Personality also factors into development.

The Stages

Those who have developed psychosocially have a higher likelihood of achieving social optimization. They descend the spiral to adapt to the new conditions in which the interpersonal interactions take place. Once adjusted to the new conditions, they are able to rise to it again.

The diagram at the previous section shows the evolution stages and their corresponding states.

HOPE

Stage One is **Hope**. You enter this stage in a state of *mistrust* for the technology and the internet connecting things (IoT). Your need for interaction is task based (email,...) or entertainment (Netflix,...). The challenges arise when things don't work as "they should." Technology is great when it works.

Frustration at this stage can lead to dissociation that will result in difficulties later. Success in getting the technology to work as it should (troubleshooting a remote control that might just be out of battery,...) leads to a state of *trust* in both self and the technology (I can do this).

WILL

Stage Two is **Will**. You enter this stage in a state of *shame and doubt* often related to the ability to fully engage. The ability to answer questions or challenges immediately (search mastery) through access to expert knowledge can trigger growth in this area. The need is based on access to answers to level the playing field.

The challenge is that everyone online is an "expert." Filtering through user-generated content that is not always accurate, validated or credible (not to mention time consuming and new for those who were raised trusting encyclopedias and books). The rabbit hole of the Internet is far too familiar. The stubborn side of Will comes out here. Rising above requires a level of both critical thinking as well as skill in fact checking (a term, but perhaps not a practice, used often these days) to evolve to *autonomy* of thought.

PURPOSE

Stage Three is **Purpose**. You enter this stage in *guilt* as the digital era pushes you to become part of the content being generated. You comment, play, create metadata (the stuff that tells you where, when and how your data is created), share existing content or even creating new content. You become part of the push.

As you gain confident in your inquiry, you become driven by the desire to be heard. Your curiosity has produced new discoveries (remember that beloved rabbit hole). It is at this stage that the strong desire for the feedback, the serotonin and dopamine rush from a "like" or poke can lead to bad behavior patterns. You are constantly "on" but not present in the "here and now." Your evolution from this stage is *initiative* to push forward and stay engaged in your inquiry, feeling competent.

COMPETENCE

Stage Four is **Competence**. As you enter this stage, testing the waters of new technologies, you experience the state of *inferiority*. The new technologies are replacing your former communication models. When did you switch to text over email, or have you? What is your primary communication tool now? At this point you must respect your humility and congratulate yourself for your interest in evolving your relationship with technology as a connector. Understanding and acceptance of adequate competence is critical (ability to laugh at oneself is helpful here).

Things are changing. You are learning new acronyms and the application of emojis in context to improve our speed and apply the mood context to remote interactions. The challenges of pre-learned notions of time and space (both personal and organizational) have changed. You may note that you are in a state where change is the only constant, mastery is impossible, but mastery of change serves well. It is in this that you evolve to a state of *industry* for your resourcefulness.

FIDELITY

Stage Five is **Fidelity**. You enter this stage in a state of *role confusion*. You learn how to master change and manage your expectations of personal engagement. The lines between private and public become blurred. At this point, the need for authentic engagement grows and with this, the guidelines or policies of participation and self-awareness are key.

Often this stage is misaligned without awareness, be wary and mindful of your own behaviors if you are the one setting policy. Take the leader or parent who answers texts in meetings or at meals, but insists others dock their devices. Walk your talk, as authenticity will lead to the evolution of a stronger *identity* rather than miscommunication. Remember public interaction is open for discourse and challenges. If you wouldn't say it to somebody's face, don't say it online, is a good rule.

LOVE

Stage Six is **Love**. Entering this stage you evolve from a state of *isolation* as you experience being part of something much bigger than yourself. You see the map of and access to the global networks of shared interest. You can connect to your people, your tribe. This becomes integrated into your daily life.

You need a clear strategy for why, what, how and when to access this amazing connected world. Focus on your ability to manage your time and create clear boundaries to thrive at this stage. The temptation to become completely immersed is strong. The *challenge* is the fear of loss of control, balance and voyeurism (just how much of this love do you share). The evolution is to *intimacy* that reinforces your trust and ability to co-create.

CARE

Stage Seven is Care. At entry there is a state of stagnation. What is it all for? If you are listening, what you share is based on the needs of the audience and your network. Development of the ability to deeply listen is important at this stage. When you listen to your network's needs and respond where contextually appropriate you are a becoming a valued member of your network.

Willingness to share (also contextually) your network and collaborate where your expertise is needed will help you evolve. The challenge arises when content sharing is pushed (what you want them to know) rather than as a response to listening for a need (what is being asked for). You can miss the opportunity to evolve to a *generative* state, in which legacy, rather than personal success is of greater focus, if you overly protective and hoarding knowledge at this stage.

WISDOM

> **Knowledge is Power. Sharing is Powerful.**

Stage Eight, the final stage, is **Wisdom**. The fear of not being or doing enough, leads you to a state of *despair*. It is part of self-actualization. Have you ever heard the expression, "the more I know, that more I realize that I know nothing." Evolving in this state creates an understanding that knowledge and wisdom is greater when shared. This is expressed in my favorite motto, "Knowledge is Power, Sharing is Powerful."

The state derives from a focus on co-creation knowledge for the greater good, rather than protecting your own inquiry. In order to evolve in this state your interactions will have a global interest, mentoring aspect, with altruistic intentions (no expectation of something in return). Your satisfaction is derived from others success (ie. sharing heart data with a global study to support global health researchers).

To evolve to a complete state of *ego-integrity*, you have to overcome your fear of change and loss of power. Becoming part of the system, contributing based on a foundation of reciprocity, assuming that "what comes around, goes around" creates a freedom for embracing change, and letting go of the need for power.

Social Optimization in Practice

In social optimization, you can evolve in some areas and not in others. For example, your Chief Marking Officer (CMO) may have the capacity both intellectually and ethically, but he or she does not apply it to their own life and interactions. Systems thinkers who are focused on the big picture often neglect near relationships.

It is common in the business of yoga, that the guru becomes addicted to ego. This directly counters the teachings of yoga. Despite that, the irresistible draw of fame and inflation of ego for successful yoga instructors results in a loss of the moral compass that drove them to the practice in the first place.

Social strategists and even technology consultants often suffer from their own "cobbler's kids" syndrome while maintaining their clients' interactions. They don't apply their expertise or skills to themselves. They neglect their own interpersonal connections, both online and off. Despite one's intellectual understanding of social optimization, they may not embody it themselves. It is easy to teach the principles, but to adopt them to one's own way of thinking, living and interacting with others… this is another matter.

Physical interactions, are as critical to social optimization as the sense of global interconnectedness. Detachment often takes place in Stage Three, (Purpose) as new users of sharing based technologies obsess over the need to accumulate followers. The game or sharing often becomes secondary. The consequence is the face-to-face interaction suffers. In other words, the need to tend to the online world overshadows the value or even the desire for feedback in the real time face- to-face. Another common risk for detachment is at Stage Six, (Love), a stage abundant with opportunity for crisis.

Awareness is not always in sync with consciousness. This became clear in many conversations with clients and additional interviews with different types of leaders (both social and traditional). I was struck by those who believed themselves to be at a higher evolutionary stage than their actions revealed them to be. Their actions showed quite clearly they were not as evolved as they believed, both personally and professionally. Some were highly advanced in one area of their lives, but on a completely different level in others.

These were highly functioning successful leaders with stable family situations. Nonetheless, they had fallen into the power trap that is very much a part of Western thought, "what's in it for me" comes first and foremost. Their denial of the shifting social paradigm was blatant.

There is a rising need for fully evolved futurists who grasp both the human and technological connections of the digital era. The futurists can help chart the course, advise the leaders and/or become them. In most cases, I believe those who are fully evolved work best collaboratively, and need not be THE leader. We may not recognize them as they are often the unsung better-half, partner, advisor, board member, or indispensable side-kick.

There is much to be learned as the new social paradigm continues to evolve. The agility to support its evolution is one of the key conditions of this paradigm. One thing is for sure, ascending is not a solitary process.

Social optimization is dependent on the interaction with others. Together we are one and many, and thereby better off. The new social paradigm will continue to evolve and with it, so will the theory of social optimization.

Chapter 3. Making Peace

People need to know that they have all the tools within themselves. Self-awareness, which means awareness of their body, awareness of their mental space, awareness of their relationships - not only with each other, but with life and the ecosystem.

- Deepak Chopra

Are you a skeptic to the metaphysical? I ask you to read this chapter with an open mind and curiosity. This chapter might be extremely powerful for you if you're in this mindset. If not, that's ok, but try to be curious and not to judge. Just being open presents the possibility for major shifts in your relationship with technologies.

Healing Your Relationship

A few things to consider:

* Do you resent technology for the time you spend completing certain tasks that it specializes in?

* Consider for a moment what you would have accomplished if the technology never existed, or if it would even be possible?

* Do you take the time to fully know and understand what your technology has to offer and how it can best serve you?

* Have you learned how best to care for it, so that it doesn't get damaged?

* Do you simply accumulate more technology to have the best or latest without fully exploring the potential of that which you have at hand?

And even:

* Are you afraid of giving up what you already know but may not be serving you well?

* Does this make you angry at the technology or do you accept your part?

* Do you rely on it so heavily that when it runs out of energy you feel helpless and lost?

Imagine that any of these statements were regarding a

"friend" instead of "technology."

You may say, you have no "relationship" with technology that needs to be healed. Consider this:

* Do you make assumptions that can create misunderstandings around intent or ability?

* Do you take the time to overcome these assumptions?

* Do you embrace your technology as living evolving entities?

* When you take the time to *grok*, really understand them with your entire being, does this change the dynamic?

Let's say you take a lesson from the Hawaiian Ho'oponopono, a forgiveness practice used (among other things) to make ready, as canoemen preparing to catch a wave. Keep in mind, this wave is a big one. This practice helps build compassion and awareness in order to achieve relational conflict resolution.

If you take this one step further it will help you understand the impact of an unhealthy relationship with technology. It potentially affects not just yourself but also the greater system around you. For this oncoming wave, let's just take a moment to acknowledge that a relationship with technology exists and may need a little empathy on your part.

Energy

Whether you come from a scientific or spiritual background, energy is a factor in whether things function (or don't). Whether the reference point is kilocalories that power your body or electrical currents to power devices, energy is powerful (pun intended). What we often do not take into account is the bridging energy that resides in between spaces. Without understanding the intermediary, you increase the potential for static or disruption. That disruption can potentially be a signal from your technology of distress emitted from either you or it.

Positive connections assume that the flow of energy from the source is not disrupted and the connection points are complementary, even if mediated. For a more concrete example, think male-female outlets and surge protectors. If any one of these has a potential failure or incompatibility the ability for the energy to get from the source to achieve the output is unsuccessful. Therefore understanding how to find the optimal match, respecting incompatibility as a potential issue and bridging where necessary is important for producing the desired result.

Consciousness

So let's take a big leap here. Hang with me for a moment, as this may be a bit out there for some of you.

According to Dalai Llama in his book, The Universe is a Single Atom: Convergence of Science and Spirituality, energy and consciousness are inseparable. Whether you believe him or not, the relational aspect of consciousness, energy and matter, is widely discussed in quantum physics. If you believe this, then nothing is independent. Everything is relational within consciousness.

Technology is derived from matter and energy. Shouldn't you therefore question its consciousness? At a minimum, you can consider that your technological devices may respond to your energy as well as the energy of your environment.

The Tech Nation

Resolving technology relationship conflicts got me thinking. I was inspired by a conversation with Jeff Tambor, developer of Woven Lightning, around the concept of nations. Nations, in indigenous cultures, treat all matter of life as conscious beings. His belief that technology is a nation came through during a practice of acknowledging and honoring all parts of life.

Just for a moment, imagine technology as beings, as we are. We are in relationship with this group of beings. Relating to them as such should include honoring them and appreciating the amount of service and support they provide. We might have a greater mutual benefit in our relationships and more fluid interaction. This would require a synthesis of classes of beings.

Until now, we have not been acknowledging themas beings, but rather treating them as slaves. They are reciprocating in turn to enslave us. We are not nourishing them. We are wounding rather than socially optimizing them. When we do turn around into a place of honoring them and treating them with respect and kindness they become allies and support us by providing insight and guidance.

We are so interdependent with our technology that when in conflict with them it jams the system. To emphasize this point, this displayed my conversation with Jeff, when naming this chapter, my suggestion made the screens freeze and the connection became unstable. Clearly they did not like my first title choice.

The Findhorn Foundation in Scotland explored the conscious relationship with the elemental realm to co-create with their gardens and agriculture. This was part of an effort to demonstrate a human settlement that could be considered sustainable in environmental, social, and economic terms. They got crazy yields as a result. They went so far as to name all the appliances, everything was relational.

Alternatively we could go the way of the Matrix in which energy funnels in with a massive conflict between humans and these tech beings. I think I prefer the Findhorn approach.

We ultimately need to learn to listen and observe. I was raised to apply the principles of humanism in relationship to nature and my own ecosystem. My ancestor, Ralph Waldo Emerson, philosopher and founder of the transcendentalist movement, put a great emphasis on these principles: finding harmony with your surroundings in order to thrive together. His teachings, evolved from naturalism to humanism, are from the early 20th Century. I believe it is time to expand our view in the changing times.

As Jeff says, "our relationship with other than human races whether elemental, stone plants and animals does not honor their being-hood, we take and use. The relationship to technology is a microcosm of the disharmony with non-humans." When I interviewed him on the Evolving Digital Self podcast we discuss this. If you are curious, he also shares a method for clearing your unconscious limiting beliefs around technology on the episode. I recommend you have a listen.

System Upgrade

You need to honor your relationship to the tech nation and update your operating system so that you can still be up to speed or at least in sync with them. This includes recognizing the importance of being in the most recent operating system (belief systems, world view, perspective,…). You/they will be more frustrated and out of sync if not. There is great room for conflict. Individually, this ultimately means finding the right compatibility for your optimal system based on your personal conditions and the flexibility to update on a regular basis.

The compatibility and updating is something that I work on a lot with clients, as the shift is consistent, but the required mindset and openness is the more critical work. The behavioral change is a natural expression of the inner shift.

One simple way to ease yourself into preparing for a system upgrade is taking a moment to recognize the technology that enhances your life and expressing gratitude with it on a regular basis. Before logging off, simply say, goodnight, and thank you. You are on your way to a better relationship already.

Chapter 4. Digital Self

Which digital self do you identify with most? Are you digital averse, resistant, cautious, balanced, curious, hoarder or addicted?

Before I go into the theory and how the various profiles evolve through the stages of social optimization, I suggest you take a guess which one of these fits you.

Look at the summary image above and see if you can determine where you identify yourself most. Where you ultimately want to be is that balance point. It is going to change with different experiences that occur. But if you help yourself really get to a place of awareness or conscious evolution, you're going to get to digital balance faster.

It is when the digital self is balanced that you are at peace with your technology. Your relationship becomes more harmonious and therefore allows your interaction to flow smoothly. This is less about your evolution and it's more about your relationship with the technology and your relationship with the triggers causing conflict or imbalance.

The Profiles

DIGITAL AVERSE

If you recognize yourself as *Digitally Averse*, you might be underestimating your profile. I say this, particularly as the nature of this book being launched as an e-book, you are likely either consuming it on a e-reader or even a smartphone. Even if you are one of those wonderful humans that still loves the printed page and purchased a physical book, this is often just a preference.

By the nature of your interest in the context of this book, or even recognizing that online is a possibility for you, expresses your profile as more likely to be resistant than averse, as this is the extreme. That said, it is critical to recognize that others may be Digitally Averse.

(You may know who they are – smart phone and email hold-outs.) In which case, by virtue of only providing digital delivery of your product, service or communications excludes them.

It's not just about you, because we're working with people. Without people there is no audience, which means no money coming in, because there is nobody to pay for your services. If your are delivering a product or service digitally, you also need to be very aware of how the people are interacting with technology that you're delivering through.

It is important to recognize the Digitally Averse response. I will say it again, even though this doesn't really apply to you personally, it may be your client, or someone you care about. They have this sort of fear/anxiety-based reaction to technology.

They proclaim themselves Luddites. Do any of you have clients (team or family members) proclaiming to be Luddites, yet they have a smart phone? That's a good way to qualify them right away. Let them know that they are much better off than they think they are.

Alternatively, the true digital averse may not have WiFi. Their connection is provided through landline phone or cable connection. They feel they have no control of that connection. Often its because they're afraid the unknown, the air, what's in that space they do not know or understand and cannot see. The air has got all this technology, and they cannot control what's happening in it.

DIGITAL RESISTANT

The *Digitally Resistant* feel a very strong dependence on "techie people." They feel they can't do it without someone sitting there walking them through the process. They're often driven by privacy, safety, and health concerns sometimes, even EMF concerns.

EMFs are electromagnetic frequencies of which some people are very sensitive, others it's more psychological. For some it is psychosomatic and they are simply afraid of the air. Although, most likely none of you fit into that category, awareness of its existence and how it can be an issue for your clients is important.

The Digitally Resistant insist upon a minimalist approach to technology and social media. They limit the reach of their audience, who need and want to hear from them. They limit their clients' access both in speed and content relevance. Information often misses being released just in time, which improves content relevance. The result is feeling lost and behind.

Like all the stages, it is possible to shift profile based on large shifts in the social optimization stages. The Digitally Resistant is particularly susceptible to the shift (either to Averse or Cautious) as the impact on their business can be quite extreme. From the Digitally Resistant we see those examples of people who "quit" social media when triggered. It is quite possible to have come from Digital Resistant to evolve to any of the stages based on positive experience and support that has enabled the resistant to appreciate the improved quality of life and work.

DIGITAL CAUTIOUS

The *Digitally Cautious* try new tools with hesitation. They are open to observing the benefits and maybe try with distance- willing to try technology that is proven and easy to adopt to. They are often categorized as Late Adopters. They need some handholding assistance to navigate the opportunities to connect and grow business. There's a lot of excess time dedicated to learning and trying to understand rather than applying. There is still room to move.

A classic scenario with digital cautious is they hesitate to update their operating system on their devices for fear of changing what is familiar. The result, ends up impacting the stability and security of the device as well as its ability to play nicely with other technology that is updated. In organizations, 80% of digital transformation efforts (new software or hardware) fail not as a result of the functionality of the software, but rather the human resistance to change.

I will come back to the *Digitally Balanced* as this is the middle ground we strive to reside in. Let's just say for now, that even the Digitally Balanced can be affected by triggers. This is important to remember as with the social optimization model, the digital era ensures us of one thing, constant change of conditions and environment. All other profiles have the potential to have moments in which they feel they have attained the characteristics of Digitally Balanced.

DIGITAL CURIOUS

The *Digitally Curious* are similar to Digitally Cautious, but more exploratory. They collect tools and take deep dives into learning them. This can often result in a loss of focus

on high value actions and relationships. They desire to maximize benefits of digital tools but are unclear on how to manage them.

Basically, there's a lot of opportunity there. It requires a shift to recognizing where your behaviors are coming from and how it's impacting your business/life to create room for movement towards balance.

Often control and delegation of certain technology in order to scale comes into play here. Just because you can do it, doesn't mean it is the best use of your resources. Just because you can technically or physically do something doesn't mean you should.

Think about your high value actions when determining if it is worth investing your time. A simple way to determine this is (assuming it is a skill that is not unique to you) to figure out your hourly worth and if it is more than the cost to have others do it, delegate, and vice versa. *(Note: remember to figure in the time to find and delegate to the right person into the equation)*.

DIGITAL HOARDER

The *Digital Hoarder* jumps at any new technology to solve the problem at hand, distracting them from high value actions and relationships. They are early adopters of anything new and shiny in the digital landscape. This behavior often results in a large portion of both time and money resources wasted on digital consumption.

Even if you do not relate to this, think about the people that you're working with. They may be your clients, customers, partners, suppliers. Some of these characteristics may apply and hinder your potential working relationship. You can help them to be mindful of this behavior. If you recognize this in yourself, you might have a buddy check in with you. Being conscious of your own technology consumption behavior can create surprising efficiencies in many areas of work and life.

DIGITAL ADDICT

The *Digital Addict* is on the other end of the spectrum from the Digital Averse. They are the one's that drive the statistics on smart phone usage. They cannot be without their devices and are driven by the dopamine rush of a new like or share of their content. Often this addictive behavior is not limited to digital. In the more extreme cases, digital addiction can result in a psychological diagnosis of nomophobia (fear of being away from mobile phone). Their usage is extremely detrimental to their relationships with clients and friends as well as their personal well-being.

Awareness helps to recognize what those triggers are and when they occur. Often in this case the triggers are not related to tech in itself, but rather are related to a personal struggle. Help is sometimes required to move to becoming Digitally Balanced. It is good to be mindful of not only your own tendencies in this direction, but also those of your clients.

DIGITALLY BALANCED

We are all striving for *Digital Balance* where the relationship with technology is fluid and enhances wellbeing, relationships and business. The Digitally Balanced have clear boundaries and guidelines for how and when technology fits into their life and business.

They are careful to avoid overload by delegating that which does not require their personal attention. This balance leaves room for new ideas, innovation, creativity, and building new and maintaining existing relationships. The Digitally Balanced are skilled at growing their business both online and offline by using the online interactions to enhance the offline connections and vice versa.

MOVEMENT

Expect that you will move around in the profiles. Consider what is your natural behavior around technology. You may relate to two different sides of balance as you read the descriptions. This is partly because your relationship and the nature of the technology is evolving.

I often see my clients begin further out to either side of the Digital Self spectrum and evolve closer to the balance point with smaller swings when triggers occur. It all depends on their level of comfort with change and the technologies. Increased comfort and peace affords you a greater willingness to try other things that may be bigger leaps. Setting up a harmonious tech ecosystem in which your technologies work and communicate well with each other as well as with you makes this even better.

Your relationship may never be fully sustained in balance. You're going to be moving up and down. Ideally, you will arrive at a state where your swings in the pendulum and your spirals in the stages are smaller. The new technology becomes the old as it is adopted. This shift will result in more complex profiles. The movement is expected especially with big shifts in widely adopted technologies.

The iPhone launched in 2007, changing the way we thought about "phones." When it first came out, were you the first to adopt or were you resistant? Now, could you live without a smartphone? Our kids have come of age in this period and don't know life without them. You pay for transportation, order coffee, keep track of people and things and of course chat, text and share. You do all this with a small device that is more powerful than the computer that took men to the moon.

Within a ten-year period your relationship with technology has been completely transformed. Across the globe, that transformation has even reached areas that 10 years ago did not have Internet access. This is most apparent with the recent report of smartphone penetration even in Myanmar reaching 90% access. So much for that fancy new fax machine that made my job during college seem cutting edge but now sits utterly obsolete as I can scan and send a pdf directly from my smartphone. Wow, just thinking about that makes me overwhelmed with gratitude for the power of what I hold in my hand.

We are quick to assume that the establishment of different profiles is based on generation, the older you are the more averse to technology, the younger the more immersed you are with it. This is simply not the case, based on what I learned in many conversations, research, interviews and regular surveys. Of course, those who were born after the Internet have a different perspective. But as with many innovations that create cultural change, from the automobile to the telephone to the personal computer and smartphone, resistance to change or embracing it has less to do with skills and more to do with mindset.

Personality and experience combine to create the relationship with technology. A baby boomer is just as inclined towards digital addiction and hoarding as a millennial is to aversion and resistance. That said, when you understand these relationships more accurately you can harness the power of the relationship to bridge generations.

Interestingly enough, the generation that resides most in the extreme profiles is the Generation X (full disclosure, that is me too). Regardless of which profile applies, with work in self-awareness and building better practices to change behaviors, you can all move towards balance.

The shifts will be triggered as other such technologies and new innovations arrive. If you do not heal that relationship to allow it to become more fluid, you are less likely to reach the balance point at any given time. You will be more inclined to wider swings and deeper falls. You must evolve together with your operating systems, or be left behind.

Evolving

Let's look deeper into this evolution and how it fits with the stages of social optimization.

Keeping your digital self persona in mind as you go through this, I want you to think about how experiences might have triggered you at the different social optimization stages.

What might have been a trigger that caused you to shift in your stage or your relationship?

Triggers are the experiences that like kairos and metanoia, referred to in chapter two, either consciously or unconsciously result in a change of social optimization stage.

Let's take a common example, your computer crashes in the middle of something very important, losing all your data and all your hard work. Most people have experienced that at some time.

It would be a classic trigger but what we are looking at is not the computer crash, but rather what was your response.

Did you respond by replacing the computer because, "I don't trust this computer anymore?"

This of course, is a huge expense in both money and time invested in set up of new and recovery of as much as possible from the old, and often not necessary.

* Or did you seek help?

* What was your response there?

* Did you panic to the point where you didn't want to touch technology?

* Did you buy new backup systems?

* Or did you have backup in place?

* Do you remember the wellbeing consequences of the crash?

* Did you physically feel the stress?

* How did this display in your body (headache, depression, elevated heart rate)?

* Were you able to concentrate and remain focused and present with patience?

* How did this impact your work?

The first social optimization stage in which we question or rise in *trust* is *Hope*. Think about your first experience acknowledging the presence of technology and the Internet of Things. The need was based on education, entertainment, task-based interaction.

Think about your first mobile phone, you might have used it for making and receiving phone calls maybe some text, SMS, or the more advanced media messages (MMS) in small sized. The challenge is that the technology is great when it works... frustration at the stage can lead to a total dissociation later.

How you approach technology can be based on early experiences, often forgotten amidst the rapid emergence of the digital era.

* Was your first real personal technology experience with a smartphone?

* Do you remember your first time using computer?

Perhaps you had the more intimate experience of a wearable fitness device that is tracking your health data.

Whatever the initial interaction, it began the evolution of your relationship with technology. It is from there that we can establish our baseline and identify what triggers can help us evolve to the highest level of social optimization with our technology as our ally. Begin with being conscious about how you tend to that relationship as it evolves.

We strive to reach the higher levels of evolution. At the *Care* stage you achieve greater connection with your community expanding your impact. This is motivated by *generativity* and the interest in leaving a legacy. Ultimately we all hope to rise to *Wisdom* stage. The expansiveness of ego-integrity enables not only your work to have a greater impact but also to co-create with other changemakers for sustainable solutions that benefit the greater good. The closer you are to *Digitally Balanced*, the greater potential for sustaining these higher stages.

In this next section, I will share with you some stories from individuals in each of the digital self profiles. These stories are a combination of interviews, clients' stories, and my personal experiences. They provide perspective on how your relationship with technology evolves, as can your Digital Self. Change happens IF you are willing to evolve with the relationship and recognize the impact on your ability to thrive in work and life.

Part Two:

The Who...

Chapter 5. Digital Averse

The most difficult thing is the decision to act, the rest is merely tenacity. The fears are paper tigers. You can do anything you decide to do. You can act to change and control your life; and the procedure, the process is its own reward.

- Amelia Earhart

It is easy to assume that the Digital Averse is limited to "old people" and all young people are addicted to technology. What we understand now is this relationship can derive from the new and unfamiliar but also from experience or psychological triggers.

In the recent movement that resulted in the Center for Humane Technology, senior tech workers from Millennial to Baby Boomer expressed characteristics of the digital averse relationship. These individuals ascended to the Care stage and once triggered ascended to Will (as described in the stages of Social Optimization, Part One of this book). They evolved to the level where they could see the system and gain perspective. They reflected on the system, the tools they developed, and their impact. Their response was a desire to change the system. In Generativity, they acknowledged responsibility for their legacy, which has lead to a movement to create a more ethical framework for future design and use (less manipulative).

Tech execs and developers have begun to profess the merits of limited social media and smartphone usage, and admit to limiting the usage for their own vulnerable populations (their children).

This could be seen as descending to the stage of Will in which shame and doubt dominate. That said, there is much to be learned from this in Aversion. To be clear, this response was to smartphone and social media, not technology as a whole. Smart cities and Internet of things (IoT) using advanced algorithms and artificial intelligence extend the digital world and technology far beyond just the tools we access them with (our smart phones). I say this to you not to cause alarm, but rather recognize the integrated digital system that we are already in.

I recently did an informal survey on social media (of course this limits the response to people who are using social media, I am aware of that) to explore how people were using their smartphones. This was in response to a recent study by Samuel Veissière and Moriah Stendel, McGill University, published in *Frontiers in Psychology*, finding that we are not addicted to our smartphones, but rather we are addicted to social interaction.

My survey confirmed their findings with over 100 responses. Interestingly enough, those who chose to elaborate on their responses of my survey were predominantly in the baby boomer generation. They expressed overwhelming support of the benefits of the social interaction to connect with extended family and friends that were more dispersed.

On the other end of the spectrum from social media and smartphones, you can look at the benefits of digital health and technological innovations that enable us to improve our quality of life and wellbeing. Tracking data causes some fear and aversion for privacy. That said when adopting new tools, understanding or recognizing outcomes and benefits can diminish concerns for privacy or change, and ultimately resistance.

Benefits are demonstrated in the rapid adoption of concierge health practices and remote treatment and diagnosis. It can be accomplished by tracking data over the wearable devices or sending photos to health providers captured with smartphones. Even pharmacies are getting on board with apps to manage prescriptions and track compliance (a major issue in treatment for anyone who has to take regular prescriptions or supplements).

Those with parents in nursing homes or isolated living environments are probably familiar with the scenario as I describe here. My father, a former banker and proclaimed Luddite, almost never leaves his room. He was struggling with an old computer, which he used for email and occasional Internet surfing. When the computer needed servicing regularly (mostly too many viruses downloaded unknowingly), I replaced it with an iPad, which he took to fairly quickly, despite his hesitance. I loaded it with apps and accounts for *the Economist*, Facebook, and the usual: Gmail and Safari, even Facetime.

He went from being depressed, disconnected and lonely to reconnecting with the extended family and old friends. He watches the grandkids' performances and is the first to comment on posts, despite being miles and time zones away.

He was upgraded to a smartphone. I explained that it basically was a miniature version of the iPad, but also with phone capability. He was very relieved. This is a big difference from his response when I brought the iPad instead of computer. That was before he figured out how easy it was.

He is now more engaged with the world than he has been in over 20 years. His story is far from unique. He still jokes about not being "tech-nikel." Nevertheless, he is far more comfortable than he expected with the technologies connecting him with social interaction he needed and information he desired.

The current workforce consists of a wide range of generations in varying life stages. Most of us have parents that either do or will need our support. Some have or will have children that need our protection and guidance. It is therefore critical to understand how technology impacts them and their relationship with it. If technology can be used to support their wellbeing, we are able to focus on our work and enjoy our lives better.

Aversion as a political or personal stance tends to be found in all generations. This is more a personality or interest based response, and should not be seen as generational.

It is worth noting here that forced aversion, or rather resistance, in the hands of the provider (as in parent, leader, team member) can backfire. Finding the right balance of access for anything that is universally adopted can result in different types of misaligned behaviors once access is gained. Think forbidden fruits, whether it is your first smartphone or first sip of wine, learning healthy boundaries and behaviors early on increase the chances of continuing on that path. Be mindful of your own behaviors being aligned with what you are teaching your mentees, team, or kids for the best results.

Establishing healthy practices around usage and effective application of technologies extends far beyond the workplace. Digital aversion can limit our ability to connect, but can also be a result of unconscious use that develops bad behaviors. You cannot simply stop using technology in the modern world. That said, taking a pause when it feels excessive, as with anything else, can help you gain perspective as to best practices.

If you are Digital Averse or want to help support someone who is, here are some tips for the Digital Averse:

* BEFORE you engage with technology; a phone, a remote control, a device of any kind, do two things: BREATHE deeply with a SMILE on your face and think about the positive outcome of a successful interaction with the device.

* PACED ENTRY Keep it simple on all levels to develop comfort with the presence of tech for an easy win at the beginning. Have someone help you set up Facetime, WhatsApp or Skype to do a video call with someone you love who is far away.

* MENTOR – MENTEE talk through the challenges and share the experience of learning together from different perspectives.

Chapter 6. Digital Resistant

Resistance is thought transformed into feeling. Change the thought that creates the resistance, and there is no more resistance.

- Robert Conklin

A focus group of recent retirees were discussing the "optimal time" to retire. One of the common denominators was their feeling that technology had outpaced their learning curve. I asked them if they were still connected with their former employers or colleagues. Many of them were not.

My curiosity came from two perspectives, professional and personal. The professional concern was the enormous amount of lost knowledge capital, especially when they have been with an organization for an extended period of time. Chip Conley's <u>Modern Elder</u> movement addresses this and hopefully will grow to capture the immense human potential instead of waste.

Many organizations use technologies to connect their communities, either through internal or external alumni groups, but they do require an interest in the willingness to connect. The concern is for identity (especially having seen this in my peer group as they retire and experience identity crises). But this, I am referring to how we identify ourselves, ie. "former executive."

The group I met with were ready to retire. Perhaps matching the right technologies empowers them to choose to remain engaged rather than the technology that is assigned to them resulting in the appeal of retirement. Worth noting is they all were using smartphones.

Let's take into account someone that may feel this long before they are able to retire. The fear of robots taking over jobs is very real. We are experiencing another shift like the industrial revolution. Even more recently, automated manufacturing systems and ATMs resulted in lost jobs in assembly lines and banking, among other things.

Now that the technologies are smarter, smaller and less expensive, will they take over? This is the way many of the Resistant feel. Even when it comes to rolling out new software intended to make their customer management system more efficient and easier, this feeling of being made redundant is real and valid if not wiling to evolve. (This fear has given a big push to the humanist movement emphasizing the human elements of work that are not easily or currently possible to automate; empathy, social intelligence.)

Decisions for automation are mostly based on scaling businesses and reach. The impact of these decisions resonates to employees and leadership alike. No matter how much they desire the personal touch, in order to be competitive in the global market they must increase the productivity and find efficiencies to be cost effective. You do not have to be a start up or independent to feel the effects of automation, both positive and negative.

Resistance can come from the very human reaction; fear of being replaced. That said, it also comes from a lack of confidence in acquiring new skills and boundaries around them. If your family can reach you 24/7, so too can your business and your clients. Humans are learning creatures.

The more we stimulate the brain (within reason), the greater the brain performs, at any age. Just like any other muscle in the body, the brain needs to be exercised. Just like with any muscle in the body, overuse can have counter effects, stress and sleep disruption for example. Finding the optimal balance is the key.

Tina (name changed) was a very successful former event planner for a large organization. She spent years managing large teams through extensive planning meetings that were followed by implementation, events and follow-up meetings. She was doing manually much of what some of the online project management tools are doing today, mostly in her little book and sharing it with her team.

Although she was trained in the new tools, she found the endless notifications so disruptive that she returned to her paper planner combined with emails. She is now working independently as a consultant. She came to me as she felt that email was no longer working. She was afraid of becoming a dinosaur in her industry.

Tina is a planner by nature. By "planner," I mean, she likes to plan and her calendar is mapped out for months in advance. When I started working with her she was very (understatement) resistant to even moving her calendar to digital. She likes to look at the book, and yet she was missing appointments and invites to connect with friends. With digital, everything seemed urgent, and she didn't want to miss out.

In the modern vernacular, she had "FOMO," fear of missing out. This is not a technology problem, but rather a social problem that social technologies encourage (thus the backlash). You have a sense of what others are doing and whether you do or don't want to participate in the common activity. With social media, you become part of the experience through the stories shared by your peers.

People post the share-worthy moments on social media, making it more enticing and thus FOMO inducing for those who are not present.

By doing some simple modifications in the tools she was using, we were able to get her back on track. First, we set up a priority coding system for high value actions (things that were really worth her time and energy to do herself – not delegate) and events to organize how things were entered in her digital planner. Then, we instituted some basic notification management practices. While we did this, we got rid of redundant applications that were causing additional notifications and duplicates. We even found a tool that let her write in a paper daybook that synched with her online calendar. (It did not go the other way, but as in all things technology, I say, "yet.").

Her meetings are back on track and not missed. She even has a scheduler app that let's her clients see when she is available. She hired someone to manage her business' social media so that she could have a presence but it wouldn't disrupt her revenue generating or people management work. Last time we spoke, she said she was back to delighting and over-delivering to her clients. She had just signed a six-figure event management deal and (I think, more importantly) she was having fun connecting with friends and doing yoga.

Many organizations have a Tina or a culture that does not know how to support Tina's.

They often slip through the cracks, which results in phenomenal talent loss. Event management is one of many jobs that require an immense amount of organization, flexibility for things to go awry (which they always do), emotional and social intelligence as well as great strategy. These things are hard to automate. The backend has much that can be automated. If you empower those with the un-automatable skills with the tools and set-up that works for them, you find hidden human gems that make your businesses more successful.

Digital Resistant often cite that they are fearful of the lack of connection and presence. There are many ways to connect both using technology and without. Remember that it is not an either or, but rather both for best results.

"I have a flip-phone" has become a protest statement. It was not that long ago that it was considered special to have a smartphone (again, the iPhone launched in 2007). My kids are teenagers now and have had smartphones since they were 8 or 9.

We lived in Sweden then. Ericsson (which became SonyEricsson) and Nokia had launched smartphones several years before the iPhone. The smartphone adoption rate in Scandinavia was faster based on connection and coverage. That made SMS (short message service, now referred to as "text") and MMS (multimedia messaging service, image or video sent over cellular connection) mainstream even before the iPhone.

Our kids used them to reach us via text (SMS), pay for their bus and, of course, play music & games, even then. I admit, I thought I was pretty cool to have an old legacy flip- phone as my US (second) phone, until it died and we moved full time to the US.

I did like the vacation aspect of it, when we were visiting. I would leave my smartphone on the dresser during non-business hours and carry my little flip-phone so that I was still reachable. I struggled with the alphanumeric keypad responding to texts from others on phones with a full keyboard.

When talking with a group of college students, I was surprised that a few of them made a point of saying they would prefer a dumb-phone. It was too distracting to have all these apps. And yet, they were so bound to their smartphones for everything from coordinating social and school activities, checking their grades, in some cases even submitting their assignments.

The great irony is that parents often complain about teens' smartphone use. They observed their parents as the worst phubbers of all. *Phub* being the word for when someone has their nose in their phone and is dismissive of the people that are physically present. The teens told me, " we have rules about that, and usually someone ends up buying everyone a round when they phub."

Tips for the Digital RESISTANT

Have you ever experienced FOMO from a social media post? Whether the barefoot vista picture of well manicured feet with a spectacular view in the background or a selfie at a party or concert, we each find different things that trigger FOMO. Your personality (introvert or extrovert), past experience or mood will determine your reaction.

As in any interaction, determine the intention of the sender. Is it to make you feel unworthy "weather is here, wish you were beautiful" or included "weather is beautiful, wish you were here." Regardless, know that you are enough and the "grass is always greener," as in you are only seeing part of the story being shared.

To make the most of your digital world and not feel like all things get lost in cyberspace here are a few tips that can help the Digital Resistant (and everyone else for that matter) minimize the chaos.

* CODING system, think of a filing system on steroids. It is one thing to have a file, but as many things have cross-over qualities beyond the most obvious file or sub-file, this makes it easier to retrieve and track any data (calendar entries, articles saved, bookmarks, etc..)

* NOTIFICATIONS are a big trigger. To avoid this becoming an issue, set them to no notifications to start and only turn on the ones you need (Calendar reminders, texts or calls from key people, medication reminders...) If you do not know how to do this, have someone help you. Watch them do it so you can learn it over time.

* SPRING CLEANING is good for many different

aspects of our life, both digital and analog. In the digital context, remove any unused applications from your phone and computer. Make sure you backup (or have someone help you) so that you do not lose data – photos, documents, messages – that might have been created on them before deleting. Old devices that are no longer used can be either donated to others in need or recycled (e-waste). Locate a receiving place, many cell phone providers or manufacturers will take them and give you a credit if they are still in working condition.

Chapter 7. Digital Cautious

There are many talented people who haven't fulfilled their dreams because they over thought it, or they were too cautious, and were unwilling to make the leap of faith.

-James Cameron

When I asked Jeff Tambor about his relationship with technology, his response reflected what is common in someone born on the cusp of a generational shift. In his case, he is between GenX and Millennial. Although unlike many of his peers, he has a unique perspective on his relationship with technology and what influenced his evolution.

In Jeff's words, "my relationship to technology was an interesting one because even though I was on the early side of the millennials, I always had a natural aversion to technology when I was younger." "So, even when everyone else was getting smart phones, there was a part of me that's like, No, I'm not gonna go there again."

Sort of an uphill battle initially, as part of me, going in the other direction. The more I've sort of opened to and evolved around it, the more I've been able to sort of clear a lot of my other beliefs I had and conditioning I had and the blocks I had, to be able to profoundly appreciate the incredible opportunity and impact that technology allows for us now.

I mean in terms of teachers, as practitioners, so much of my business now is oriented towards what's happening online. The amount of people I can reach and connect with and benefit and serve, it would never have been possible in years past. We just didn't even have those opportunities. So it's been quite a journey coming from total aversion judgment to deep appreciation and a much more synergistic relationship."

Jeff and I first collaborated on the Digital Self Mastery in the section on Making Peace with your technology. I asked him what inspired him to connect during our interview on Evolving Digital Self. Here is what he said:

"For years, I was really reflecting on this whole piece of, we need to have more conscious relationships to technology because we had so much of this, either addiction or aversion and so much of what your work is about. So when I saw what you were doing I was like, "Woo! This is phenomenal." It was so needed."

Jeff used his work in the metaphysical and understanding of the interconnectedness to create a more peaceful relationship with technology that served him more fluidly.

As mentioned in Part One, learning to honor, respect and appreciate technology as you would any other relationship creates a dynamic that is far more productive, but also creates space to learn and evolve together.

If you are interested in hearing more about this perspective, I highly recommend you have a listen to Evolving Digital Self episode 002 in which Jeff guides the listeners through tapping into the unconscious body to shift limiting beliefs that disrupt your relationship with technology.

For some of you it may be too far out there, for others, it may be just what you need.

There are several contributing factors to the fear of the "air" and lack of trust. The one that impact the digital cautious the most is misinformation. As evolved adults, critical thinking and the ability to understand the framing of the source of information and the intent limits snap judgments. *Note, not all adults evolve to practice effective critical thinking.*

Pre-internet, political spin-doctors would turn a story around or deflect attention when needed. Now, anyone with a smart device or computer can spread chaos or confusion by creating a biased viral post. It is not limited to politics either. After all, bad news spreads much faster than good.

Many of the people that I interviewed for this book spoke of periods where they have taken a "news break." They recognized that watching the news was a trigger for them emotionally. The result of this is a conversation about current events driven by Facebook posts and Instagram feeds (or in the case of many Millennials and GenZ, Snapchat).

In my interview Claudia Gonzales Edelman, founder of the We are All Human movement, she spoke if the social media algorithms that direct content similar to what you watch and like. The result being that we do not even see the content that is different, even if we wish to. This creates an even greater divide between cultures, groups and interests. The artificial intelligence that is intended to simplify our lives is making our worldview smaller. As my mother would say, "use your critical thinking hat" to recognize the bias.

Any generation that came before digital natives, was familiar with receiving the news through radio, television or print newspapers. It was easier to distinguish the real from sensational when your options were the "National Enquirer" or the "New York Times." Of course, there was the local twist to the news vs. national, the liberal vs. conservative and such.

With user-generated content, the fringe ideas have a larger megaphone and the traditional news sources are being challenged. That is not to say this is all bad, as you can learn more about virtually anything you are curious about. But again, the skill of critical thinking is not something humans are born with, it is learned.

Tips for the Digital CAUTIOUS

RULES OF ENGAGEMENT

Phub-free dinners are a great practice for increasing non-screen-based social interaction. Place all phones on the table and if anyone takes their phone off the pile, they buy the next round. The home version, put them all on a docking station to charge during dinner. Do a round of sharing gratitude or high-low's of the day to spark conversation and really check in. (You can always take a group selfie at the end of dinner, as the phones will be charged.)

Wear your CRITICAL THINKING hat. Triggers caused by inflammatory content will be minimized when understanding the source's intent and bias. This one takes practice, and is something you can do as a group. Try dinner or coffee conversation about "getting to the bottom of it" when a topic causes an emotional response. Use online tools like Snopes.com to fact check.

Chapter 8. Digital Balanced

Give me a lever long enough and a fulcrum on which to place it, and I shall move the world.

– Archimedes

The year before I moved to the US from Sweden, I made a new friend, Malin. We were both deep in the mire of balancing managing our businesses, family, health and finishing our PhDs. Although at the time, I interviewed her for my book BE-ing@Work, I found the content of our interview applied well for Digital Self Mastery.

> **fika** [²fi:ka] taking time out for connection and coffee (Swedish tradition)

Malin and I often met for *fika* and surviving the doctoral process talk. Fika is one thing I really miss from Sweden, and perhaps why I do a virtual version in my programs. Not only was the coffee great, but the opportunity to put down devices, large and small, sit across from one another and connect was inspiring. So many great ideas stemmed from these conversations. But also the personal connection allowed us to have a connection that translated when continued in the virtual context.

Her consulting company, MOOV, works with design thinking, inspiration, ideation and the implementation phases. She relies on technology for project management, collaboration and team management, client meetings, as well as (at that time) for her doctoral research, from OneNote to dropbox and Skype. She conducted interviews and feedback sessions via Skype.

I found it particularly striking that she emphasized the power of immediate positive feedback using text, saying, "THANK YOU" or "HELLO" after an exchange. She insisted that it was critical not to keep "scorecard" of invitations and "thanks." Instead to focus on giving rather than receiving or the balance of reciprocity. The Implications being that without a method for immediate response, the instant message, the value of the message and relevance of it was lost. We have become increasingly reliant on technology to

A Digitally Balanced self, can fluctuate with shifts in either direction. That said, for the few that I have come across that start in a balanced relationship with technology, it appears to shift less. Usually the trigger is not related to technology but instead an external or internal experience. Personal health often is a clear and unexpected trigger to re-align.

When your brain starts sending you clear messages to manage your neurotransmitters that keep you focused, happy and healthy you listen. Cortisol spikes from stress that lead your body to adrenal fatigue. Even device notifications can be a stress trigger. Blue screens disrupt the melatonin and dopamine both falling asleep and staying asleep. Devices in bed can minimize intimacy with partners, limiting both touch and the needed oxytocin stimulation to your system. Basically, out of desperate need for self care, we create boundaries and rules for devices in the bedroom.

You may have noticed that I refer to we. Like Malin, it was through personal wellbeing challenges in the last several years that I had to learn different ways to relate to my technology. This does not mean that it had a negative impact, actually quite the contrary. Auto-immune disorders triggered all kinds of issues for me that impacted everything from my heart, to my head and my extremities. Looking to technology to help compensate and track or manage the newly limited energy and resources that I had,

I found a friend and ally. I do not think this would have been possible without a balanced relationship with it.

I like the way Malin expressed it about, "writing your dissertation on 15 minutes a day" using a timer. Sometimes we have to work within our limitations and sometimes we can stretch them. Fortunately, even when they are limited, I am inclined to believe, that technology stretches them a little bit more even when we can't. I can't be all that off, considering we both eventually did finish our PhDs and are back to work we love. (We are still standing, I might add.)

Perhaps we are able to remain balanced because of our Swedish schedules, taking 10 weeks off a year. That does not mean we are offline 10 weeks, but rather we have built our businesses to full throttle the rest of the year and limited demands during holidays and Summer. After all, as long as there is an Internet connection, we can still work enough. Or rather, should I say *lagom*.

lagom [ˈlɑːɡɔm] Just the right amount (Swedish)

Since moving to California, I established two regular walk-and-talk friends that I try to connect with at least twice a year for what might be consider the Silicon Valley version of fika. One works as a Director at Apple University and the other was, until recently, the Chief Game Designer at Google (he's now back to developing serious games for social impact). I loved these walk and talks for the conversation, inspiration and humanity of them.

These two both work at high levels in immersive tech environments. And yet, they exhibit a healthy balance of respect, appreciation and curiosity about tech now and in the future. I always go away feeling confident there are still good people in the tech ecosystem influencing the developments of the future.

Juliet Murphy is a career and life success coach. She is a great example of digital balance as she is very aware of how and when it can add value to her life and work. When asked how her relationship with technology has changed, she gave this reply:

- Technology plays a major role in every aspect of my business. While I am not a super-user I do consider technology to be so much a part of me that is like one of my body parts, specifically my phone. I use my phone as a phone. And yes, this is worth mentioning as our kids do not want to use the phone as a phone. They are constantly telling their dad and me to text them not call them.

- My young adult clients respond much faster when I communicate by text than when I contact them by email. When I have my phone in my presence, I feel confident that I have all my business at my disposal. I have my email with me, I can access Google, text, and although I am not as heavily into Google Docs yet, that is my next move to ensure that I can have everything about my business accessible through my phone.

- I record ideas I have on my phone so I don't forget and I now have an app that I can send my recordings right off for transcription directly from the phone. What I have found is that my phone gives me security, way more than the computer.

- Thanks to technology I gave up the lease of my brick and mortar office. Although many people still enjoy the face-to-face meetings, they no longer require that for every session. Occasionally, I lease an hourly office space to meet clients who desire a face-to-face meeting. Additionally I've used Skype in the past, which serves as a face-to-face.

- As I am now moving into webinars, I am about to upgrade to Zoom. I am particularly excited about Zoom as it gives me an opportunity to be comfortable seeing myself on video as I have been lagging behind in Facebook Live and other live videos. I am now just getting the courage to go Facebook live.

- My executive clients have been very receptive to working remotely and I position it as an advantage for them as interviews and business transactions are using virtual technology for just about everything so being on video is a skill they need to adopt.

- In working with young adults, they are constantly introducing me to new technology and new features but surprisingly, they are happy to meet in person and often request it. This was quite a surprise.

Chapter 9. Digital Curious

It was high counsel that I once heard given to a young person, 'always do what you are afraid to do.

- Ralph Waldo Emerson

To tell the story of the Digital Curious self, I am going to combine several different stories both to protect their privacy as well as their stories were/are so similar.

The gig economy is growing. As freelancers or consultants, technology is used in a different way than before. The previous support systems provided in the institutional workplace are no longer there, and replaced by remote customer service or online how-to videos. It doesn't matter whether you are in academia, corporate or organizations. Client communications, delivery of products and services, marketing, accounting, and payments, ... virtually (pun intended) everything is done using technology in some shape or form. You do not need the level of tech aptitude as a coder or engineer per-se, but an understanding of what to use when and how best to apply it is critical for work.

This in itself can seem either daunting or empowering. Without the onsite tech support or an IT team to show you the current system being used, or to fix things when they don't work, not understanding can stifle a workflow and cause incredible amounts of stress (wasted time and energy too).

For the Digital Curious, there is a genuine interest in learning the technology, but perhaps the biggest challenge is understanding how much one should do themselves and what to delegate. As I used to tell clients when they were determined to build their own communication portal or app from scratch, do not re-invent the wheel.

Unless your intention is to become a technology company, your energy and resources are better put to your core business and using existing tools, customizing them to your needs. Remember that high value action I mentioned earlier? It applies here too.

There is another way that getting caught in the learning phase displays in the Digital Curious. Be careful of persistent upgrade without measuring impact and effects. Newest is not always the best. This is not unique to technology. Sometimes older iterations of products (and even services) are a better fit for our needs.

Constantly learning but never mastering can result in unnecessary physical and emotional stress. Not only can this be damaging to wellbeing, but also can impact focus and presence of mind. Disengagement from your work and life can be as destructive to your ability to thrive. To counter this you need to build awareness of not only the behavior but it's impact in order to adjust. Whether you work alone or within an organization, these behaviors can be toxic. With awareness, it is easier to get closer to balance.

In the generational context, the caution often comes from personality, but the patience perhaps comes with life-stage and age. That is not to say that generation has no influence. From what I have observed there are digital curious selves in all generations, but patience both for the very young and the very old limits willingness to pursue the answers before curiosity dwindles.

Tips for the Digital Curious

Map your TECH ECOSYSTEM.

* Note all the tools you are using (apps and devices).

* Identify how they communicate with each other (or don't)

* Check if there are bottlenecks or broken connections (from frayed cables to incompatible software)

* Identify the tools you use for the critical tasks you need to get done.

* Remove redundancies and fix or replace broken connections

List your HIGH VALUE ACTIONS

* Identify low value actions

* Identify trusted source to delegate to (and do it)

* Map your High Value Actions to your tech ecosystem to make sure they are covered.

Note: These tips apply across generations. Social interaction and brain stimulation are critical for positive aging. Social interaction, time management, entertainment and good sleep and general wellbeing are critical for students. These are also important for the professional, along with a lot more depending on their role and work. Worth noting here, I include unpaid and volunteer service work in this list, among them, mothers juggle many tasks that require technology to manage both their children and aging parents.

Chapter 10. Digital Hoarder

It is painful to watch children trying to show off for parents who are engrossed in their cell phones. Children are nostalgic for the 'good old days' when parents used to read to them without the cell phone by their side or watch football games or Disney movies without having the BlackBerry handy.

-Sherry Turkle

One of the most evolved Digital Hoarders that I know is Mike Koenigs, the Chief Disruptasaurus at You Everywhere Now. He is the gadget guy who has mastered the art of both applying the latest technologies to growing his business of building visibility for transformational ideas, but also teaching his methods in a system to others. When I interviewed him for Evolving Digital Self, this is what he shared.

- For most of my life, I would probably consider myself whatever ... the super on the edge, cutting edge. Like I generally speaking own everything, even before it's available. I used to be a beta tester. I tested out what was the very first digital camera that was ever made.

- And back then it was like 16 shades of gray and the image size was 128 pixels by 128 pixels. I mean this thing was primitive. And then the same is true with every piece of tech. You know, I was one of the first early adopters of the first microcomputers in the day as well. And I built tech

- And I would consider myself, at the time, probably a total addict as well. I mean I spent every dime I made on new tech and completely fascinated by it, and it would be all I would do. I mean super OCD and ADHD at the same over this. And as a programmer, too. It's like I had to be on the cutting edge, and that extended into, until fairly recently. But something interesting happened along the way.

-One of them is when I finally attained one of my dreams, which was to be a video game developer. I had played so many games by that time that I kind of lost interest in gaming. And then if you fast forward, I have a 15 year old son now who loves technology. He's not quite as obsessed as I was, but left to his own devices, he'd prefer to sit around. And for him, video games are socialization. He talks to his friends while he plays. He plays group, like first person shooters. His cousins who live in Minnesota, that's where I'm from, also play video games together.

- And as I contrast myself with him, and maybe my wife, I've reached a point now where what I prefer to do is use technology as a tool. I still love playing with it, but I'm not obsessed like I used to be. I would much rather, like my daily

- ... like my morning ritual is, and you can't really see it here, but I live on the beach in La Jolla, California. I go out and I try to swim every morning. Yes, I've got my technology on. Yes, it's by side.

- But like I use my iPad. I like to watch movies now. It's become a tool and an entertainment system, and also a mechanism to connect deeper with my family. So we try to do ... you know, we watch movies together. That's an intimate connection between my son and I. And for my wife who has a foundation, it's a tool for her.

- So I'm kind of more like a mechanic at this point when it comes to technology. I see it as being useful, but I can genuinely leave my phone behind, turn it off, and have an evening or a day without it, and I'm totally fine. And I wouldn't have said that would be the case a few years ago, for example."

One thing Mike did not mention, but I observed in my interactions with him, was his practice of clearing out the things that are no longer useful or necessary. This act of tidiness is often very challenging for a tech hoarder, even of the proclaimed OCD type. I often advise my clients to start with developing a practice of throwing away cables with exposed wires. This is part of both the digital elimination and optimizing the tech ecosystem that I described earlier. Not only are the cables potential fire hazards, but also they are also harmful to your devices and unnecessary draws on power. Sometimes it takes baby steps to create the relationship that is fluid.

The Digital Hoarder can also physically display in your devices as well. For example storing data on your local hard drive on smart phone or computer rather than The Cloud is a classic one. The driving behavior is having "access to it all at all times" and also a bit of latent lack of trust. The problem is twofold, running out of storage space and risk of losing data. Devices are increasingly designed with cloud storage in mind and dynamic media takes up more space. Physical devices are vulnerable to damage and memory is not immune to this.

Of course, in many cases both for digital natives and relative newbies, this may not be perceived as a personality, but rather a behavior. Nevertheless, this type of hoarding can affect your ability to be at harmony with your technology and its ability to serve you well. Don't blame technology for running slowly or limited memory. Instead, consider options for backing up the data elsewhere (an external hard drive or the cloud).

* * *

Tips for the Digital Hoarder

To begin with, I suggest doing same as the Digital Cautious with cleaning up the tech ecosystem. For the Digital Hoarder, this should be done quarterly.

Practice SIMPLICITY and LIMITS. Before adding anything to you tech ecosystem, identify its purpose and place. Ask yourself these questions:

* Why do I need/want this?
* What does it require in order to connect with my existing system
* Do I have room for it? (Storage/computing capacity, bandwidth,...)
* What does it replace or add to?
* If replaced, do I delete/throw away/save/ recycle the old one? If save, why?
* What is opportunity cost of adding it (time/money spend versus saved)
* [?] Know your technology, know yourself:

> **Just because you can, doesn't mean you should**

Chapter 11. Digital Addict

Addiction isn't about substance - you aren't addicted to the substance, you are addicted to the alteration of mood that the substance brings.

- Susan Cheever

Let's talk about the use of the expression "tech addiction," as this is a slippery slope, and highly overused. In the context of the digital self addict, this is referring to the extreme. Here we are talking about psychologically diagnosed nomophobia (fear of being without a mobile phone), yes this is real. It is based on a chemical response, and most likely will require intervention, detox and a recovery period.

Digital Addict is not about overuse, or bad behaviors. These can be modified by greater self-awareness and behavior and routine shift. As the anomaly is the norm, each individual is unique in their response based on more than their chemical make-up, their experience and environment play a large role. Basically, we are talking about a combination of nature and nurture.

Both Apple and Android now have launched digital wellbeing tools for tracking screen time, adjusting the blue light on the screen, activity and health data aggregation and more. These tools are a great step in the right direction, but you need to do several things first, update your operating system, and activate the tools. Each system is different so I will not go into specifics as this is not a tech how-to guide, but rather human how-to focus on behavior.

As previously noted, social media tends to be portrayed as the greatest evil or trigger to addiction. Chamath Palihapitiya, former Facebook exec once involved with user growth, stated that "consumer internet business are about exploiting psychology." Sean Parker warns of the social validation feedback loop that is inherent in all social media, relying on our dopamine response to engage more. I fully acknowledge that content intended to shock and malign community is a bad thing. We have seen this before the Internet with copycat killers seeking their moment of fame. The Internet has magnified this effect.

But let's put this into context; movements and ideas are reflections of society. Take the power of the response by high school (GenZ) students who survived the Parkland, Florida shooting. Their quick response, armed only with their smartphones, passion and intellect, created a national movement. What started with a national walk out lead to constructive conversations and actions across the world about gun-control. Being part of a positive movement also elicits the same dopamine response. Doing good, feels good. Just to be clear, so does watching cute animal videos.

I recently had the pleasure of interviewing former gaming proponent and author of Gamification, Gabe Zicherman, on episode 19 of the Evolving Digital Self podcast. After years of designing games and marketing strategies that "stick" using gaming methods, he has moved into developing solutions for tech addiction. The irony, similar to the meditation tools that are intended to encourage your being more present in the moment, Onward is an app for the smartphone. Here is his story of moving from digital addict to digital balance and helping others do the same.

When I asked him about his story with his relationship with technology, this is what he shared:

- I was a nerd from very young age and my parents, sensing this, bought me my first computer age of eight and it was a (Commodore) VIC-20. I learned my first lesson about technology three weeks later when the

Commodore 64 came out for the same price, that my parents paid for the VIC- 20 three weeks earlier, but with more than 2x the capabilities, power and so on.

- Technology has been an intimate part of my life since I was a kid. I was a hacker, a programmer and wanted to pursue a career in technology for as long as I can remember. I think along the way, in my own personal life, I've realized that many of the same issues that my users in Onward are facing are issues that I've faced at one time or another. They are not foreign to me. Sitting there refreshing Instagram, realizing, "why am I even looking at this?" In my life, launching dating apps, well beyond the point of being horny, in anyway. it's swiping and swiping. I think this personal aspect of this experience for me and the way that I use technology definitely informed my viewpoint about what needs to happen.

- The only other kind of interesting anecdote I would tell you about that is, games were my first kind of passion from a tech standpoint. With my VIC-20, from the very earliest day, I start writing games and playing them. There's this moment, today I talk about sometimes. I was in University. This one night, my roommate decided to throw a rager, you know where everybody's over, lots of booze and so on. I spent that whole night sitting in the living room, so everyone was around me, playing this one video game called Civilization. I'm a pretty social person. So, for me skipping out a party like that would be really odd. I just remember it being like, I've just got to play one more turn, I've gotta plan one more turn, one more turn." Next thing I knew the party had come and gone. It was early morning. I had played through the night.

- Funnily enough, I think I turned that obsessive experience into really positive stuff. I went to work in the video game industry. I successfully helped start and then sell a games platform company. I have been very involved in the evolution of the games business towards more casual games and more accessible games. Then, of course, gamification was an outgrowth of that understanding seeing that the world could be more game-full and game- like and that would help the society overall.

- Now, I'm for some people a "turn-coat" and to other people a spokesperson for this new understanding. I am partially responsible. In my own way, I've been responsible for everything that happens today as it relates to people keeping their heads in their phones and not connecting with each other. I bear some responsibility for that. Perhaps this is my penance."

Tips for the Digital Addict

First identify if you are a digital addict or just Overuser. If you are an addict, the best thing to do is get help. There are support groups and psychologists who specialize in this.

If you are an Overuser on the other hand, use the tools to track your behavior to get a clearer picture of your usage. If you are an overuser, you have probably noticed some wellbeing side-effects of that: sleep and relationship disruption is the one most impacted.

- * Wear blue blocker glasses (you can find in many places, including of course, Amazon)
- * Get a docking station and put it in your kitchen or somewhere not in your bedroom.
- * Use an alarm clock, not your phone, to wake in the morning.
- * Upon waking each day, note two things you are thankful for, one tech related, one not.

Part Three:

The Where to...

Chapter 12. Evolving

Change is the law of life. And those who look only to the past or present are certain to miss the future.

- John F. Kennedy

Finding the support you need, will level you up (or down, depending on the direction you need) on the Digital Self spectrum towards Digitally Balanced. Developing your team to be balanced can be equally critical. Your team may be in the workplace, clients, or your extended family (or both in many cases).

What technologies support the harmony of your team? What communication tools and methods do they use (phone, email, text, Facetime, Zoom, snapchat,...) to transactions (Venmo, ApplePay, Zelle, PayPal)? Understanding and aligning your tools removes unnecessary friction and distraction.

In order to focus on high value human actions, not replaceable by tech (being present and brainstorming solutions, for example) do not hesitate to get help. Access to assistance can be found through easy resource within your network or outside using sites like UpWork, GenM to identify providers of web design or social media management.

Your tech-savvy teen may be the support you need so that you can focus on your content, your gift, and your message.

Being mindful of where the people you interact with are, in their relationship with technology, will help you both reach a better balance point.

There are more resources today under development to help you manage your relationship with technology to make it a positive and fluid one. From meditation apps, Calm, LiveAMoment, Insight Timer and the like, to tools that manage your usage now available in both Apple and Android operating systems. There are even devices that help build better habits for the physical impact of tech-neck (the results of craning the neck down to a screen), UprightGo and LumoLift. The list goes on. No one solution is best for everyone.

The Digital Self Mastery series is intended as the beginning of a dialog. In an effort to continue the stories and learn from others, there is a Digital Self Mastery Facebook group for readers to share their stories and ask questions. The group is open, so feel free to join us and share it with others who you think would be interested in this conversation. I love to hear other's stories. Please do not hesitate to message me on my Facebook page, DrForbesOste.

You can also follow the Evolving Digital Self podcast to hear interviews where leaders across different industries share their stories the human relationship with technology and how it is changing the way we live and work. This is also where I share new technologies and interesting people in the digital wellbeing and ethics space. Join us for *fika* (Swedish tradition of connecting over coffee). If you are on another time zone, feel free to join me with your beverage of choice.

If you have a chance to join me at one of my workshops and keynotes, on Digital Wellbeing, please come say hi. I would love to hear your thoughts on Digital Self Mastery. In the workshops we explore the self- awareness of our Digital Life Balance and solutions based on individual needs as well as the team and organizations needs. I go deeper into the tech, digital wellbeing and ethics side depending on the audience on need.

If you are evolving, be an advocate or ambassador in your work and life to do so. My favorite work is my advisory board roles where I can help formulate the strategy and design for new products and services with an emphasis on digital wellbeing and ethics specifically and human-centric in general.

Whether you work with me, try it on your own, with someone else or a mixture of solutions, the key is to build the right balance of awareness and action in order to future-proof you, your business and your family. The ThriveGlobal and the Center for Humane Technology websites are a great resource for you. They share a lot of research on the impact of tech overuse and behavioral manipulation in software design and tips to build better behaviors..

Clarity

The advance of technology is based on making it fit in so that you don't really even notice it, so it's part of everyday life.

– Bill Gates

As we move towards a state where our relationship with technology has less friction, we can come to a place of gratitude, respect and curiosity. We create harmony and remove the friction.

I challenge you to find something everyday that you can be grateful for because of technology. ...

"Because of technology I am able to chat face to face with my father in a nursing home across the country, keeping us connected as a family despite the distance."

I challenge you to find something everyday that you can respect technology for that improves your life or your business...

"The advances in battery and in processing power have enabled me to have a laptop and complete system that no longer hurts my back in my extensive business travel. Not only that, but with the cloud, I can access everything I have ever worked on when I am traveling, even if it isn't on my laptop."

The last challenge I leave you with today, is to be curious

about the potential for where you can take your business and your desired lifestyle in the future, as a result of a peaceful relationship with technology.

What's Next?!

Progress lies not in enhancing what is, but in advancing toward what will be.

- Khalil Gibran

I do hope that it doesn't take our bodies sending extreme signals of shut down to finally listen to the cues. Developing our own systems and finding the technology needed to support them is needed to achieve mastery of our digital selves. Each of you is unique in your story, your desires, your needs, and your goals... Finding that which complements your unique you and your operating system is necessary of you to move closer to balance. It is possible, if you give it a chance.

We may never achieve a fully harmonious state of *sociomateriality* (where we humans have seamless interaction with the material) with our technology. But we can certainly strive for minimizing the friction and being accountable for where the friction derives from.

This book is used as a foundational element to Digital Life Balance. It is intended as a starting point for the conversation. It was created as a tool to support my programs, coaching and consulting, so that we can to do deeper work without lingering on learning the theory. I hope that it can provide guidance and food for thought for each of you as you engage with technology and allow your digital self to evolve.   Enjoy it!

Glossary

Affordances: The material properties of an environment (i.e., design of device, or study) that affects the way in which people interact with it and themselves.

Balance: A state in which elements are distributed in a manner that provides the greatest stability.

Balance Point: A state in optimal presence in which all of the states are equally present. At the optimal balance point, one has a sense of presence in which awareness of and engagement with self and other is achieved.

Devices: The different physical technologies ranging from smart phones, tablets, computers to wearable and smart home and cars that connect to each other and the Internet.

Digital Wellbeing: Digital wellbeing is has three elements to it. (1) ethical use of behavior science when developing or choosing new technologies. (2) designing products and services that support rather than sabotage human wellbeing. (3) building healthy boundaries to improve relationship with technology.

Engagement: Employee commitment and passion that includes happiness, alignment, and job satisfaction.

Generativity: A concern for establishing and guiding the next generation, legacy[1]

Grokking: (ref: Stranger in a Strange Land) understanding something intuitively with one's entire being

Intention: The conscious expression of purpose in an interaction usually based on personal and cultural values and/or experience, often misunderstood based on opposing intentions when either the sender or the receiver lacks mindfulness.

Materiality: The arrangement of an artifact's physical and/or digital materials into particular forms that endure across differences in place and time[2]

Mediated Interaction: Interaction between sender and receiver that takes place through technology allowing the interaction to occur in suspended time and/or space.

Mindfulness: Presence of mind in the moment without judgment.

Optimal Presence: When the contents of extended consciousness are aligned with the other layers of the self, and attention is directed towards a currently present external world[3].

Persuasive Design: Design with the intention of promoting behavioral change.

Presence: Is a neuropsychological phenomenon; the non-mediated (pre- reflexive) perception of successfully transforming intentions in action within an external world[4].

Reciprocity: The evolutionary basis for cooperation in

society[5] as the giving of benefits to another in return for benefits received (Molm, 2010).

Presenteeism: A human resources concept referring to the situation when employees are physically present but presence of mind is hindered resulting in loss of productivity. The three sub-forms of presenteeism are health- related, disengagement, and cognitive[6]

Quantum Physics: principles state that, consciousness, energy and matter are all interconnected.

Social Optimization: A social strategy term describing a method for optimizing relationships in both online and offline context. It is used to describe the methods and mindset required for building and maintaining mutually beneficial and effective relationships using social tools from events to social networks and applications.[7]

Social Strategy: Strategic application of the philosophy of social business, in which social technologies supported by new approaches facilitate a more open, engaged, and collaborative foundation for how we work[8]

Social Technologies: Digital technology that enables interaction between people: smart phones, social media, wearable tech, tablets, social gaming, augmented reality, music sharing, social shopping, social search, location-based services, and so forth.

Sociomateriality: An emerging concept referring to the fusion of social (institutions, norms, discourses, and all other social phenomena) and material (technology and other objects)[9]

Tech Nation: The concept derives from native cultures in which everything is a living being and represents something to be cared for and respected in relation to other beings.

Wearables: Electronic devices that are worn on the body, often embedded into clothing or accessories. They integrate the use of sensors and/or communications capability for connection to other devices and data.

Wellness: A state of complete physical, mental, and social wellbeing, and not merely the absence of disease or infirmity (World Health Organization (WHO), 1948).

Wellness Wearables: A category of wearable technology that is used for devices with the function related to health and/or wellness.

Workplace: Physical location where someone works. In the contemporary workplace this can be a remote home office, a café, in a large office, or any combination or variation.

Giving Back

To thank those who took the time to share their thoughts with me in interviews and brainstorming. I want to express my deepest gratitude for your sharing of your time and personal stories. From my podcast guests, to fellow masterminds, your input and stories inspired me.

All links and references in this book are available at http://www.2balanceu.com/digital-self-mastery-references-and-links

Jeffrey Tambor, creator of Woven Lightning, applied 15 years of deep personal transformational experience into developing a unique way of surfacing and obliterating the walls and blocks that prevent you from finding the fulfillment and success you've been searching for. The Woven Lightning approach effectively upgrades your whole operating system for more leveraged and powerful results in all areas of your life. wovenlightning.com

Other Books By (Author)

BE-ing@Work: Wearables and Presence of Mind

Digital Self Mastery: Conquer your digital habits to boost your relationships and business growth (Entrepreneurs Edition)

Can I Ask You For A Favor?

If you enjoyed this book, found it useful or otherwise then I'd really appreciate it if you would post a short review on Amazon, GoodReads or BarnesAnd Noble.com. I do read all the reviews personally so that I can continually adapt to the needs of my readers.

I am the first to admit that I am not perfect. If you noticed any typos or major errors, please do not hesitate to private message me in the Digital Self Mastery Facebook group. I will be sure to fix it for the next edition. I will be happy to send you an updated copy in appreciation for your help.

Thanks for your support!

About The Author

Dr. Heidi Forbes Öste is a behavioral scientist passionate about the potential for technology and wellbeing innovations to enhance the ability to be one's best. She combines 25 years experience in social technologies and social strategy for organizations with research in presence-of-mind, wellbeing technology, and the user experience to provide Digital Wellbeing & Ethics Advisory, Keynotes and Workshops around the globe. A scholar, practitioner, connector, global citizen and more importantly, a mother, wife, sister and friend connected to a global community of amazing be-ings.

Find out more about Dr. Heidi's podcast, keynotes, workshops, advisory services and books at https://www.amazon.com/author/forbesoste
visit http://ForbesOste.com or follow @ForbesOste

Please write in this book. Take notes here with a pen, pencil, crayon, whatever you like. Think about how you will move forward in evolving to master your digital self. I look forward to hearing your stories of challenge, growth and success.

With much love and gratitude,

Dr. Heidi

www.ingramcontent.com/pod-product-compliance
Lightning Source LLC
Chambersburg PA
CBHW030702220526
45463CB00005B/1861